EINSTEIN'S LEGACY: REVOLUTION OF RELATIVITY

THE BRIEF HISTORY

BY AMAZIR AMANATH

Einstein's Legacy:
The Revolution of Relativity

By:
Amazir Amanath
(Aspiring Astronaut
Dip. in Celestrial Navigation)

Words of Writer

Embarking on the exhilarating journey of adolescence, at a mere 15 years old, I find myself at the intersection of dreams and discovery. An unwavering passion for the cosmos propels me forward as an aspiring astronaut and devoted physics enthusiast. In the tapestry of my young life, I proudly weave the narrative of "Einstein's Legacy: Revolution of Relativity," a testament to my insatiable curiosity and eagerness to unravel the profound mysteries of the universe.

Despite my youth, the pages of this book bear witness to my earnest commitment to understanding the complexities of physics. With each word, I navigate the intricacies of Einstein's revolutionary theories, translating the esoteric language of relativity into a narrative that transcends age barriers. It is a labor of love and a testament to my determination to bridge the gap between the profound and the accessible.

Crafting "Einstein's Legacy" has been a journey of self-discovery, an odyssey into the heart of theoretical physics that has both challenged and inspired me. The ambition to comprehend the implications of Einstein's theories has not only fueled my intellectual growth but has also ignited a fervent desire to contribute to the discourse surrounding the cosmos.

In the constellation of my aspirations, the dream of becoming an astronaut shines brightly. The yearning to explore the cosmos, to transcend earthly boundaries, drives me to pursue knowledge with an unwavering dedication. My story is one of a young mind reaching for the stars, fueled by an unquenchable thirst for understanding and a determination to leave an indelible mark on the scientific landscape.

As I pen my own narrative, I hope to inspire others to embrace the boundless possibilities that lie within the pursuit of knowledge. In the vast expanse of the universe, age is but a fleeting factor, and the journey of self-discovery knows no bounds.

Table of Contents

Chapter 1 - The Revolution

1.1 - Overview of Einstein's Impact

Albert Einstein, a name synonymous with genius and scientific revolution, continues to captivate us even a century after his groundbreaking work transformed our understanding of the universe. His impact extends far beyond the realm of physics, influencing fields like cosmology, philosophy, and even popular culture. This section dives into the heart of Einstein's monumental contributions to science, exploring the core principles of his most influential theories and their enduring legacy.

1.11 - The Theory of Relativity:

Einstein's theory of relativity stands as a cornerstone of modern physics, fundamentally changing our perception of space, time, and gravity. Contrary to the established Newtonian framework, which viewed space and time as absolute and unchanging, Einstein proposed a more nuanced picture.

1.12 - Special Relativity:

This theory, published in 1905, emerged from Einstein's thought experiments and the famous Michelson-Morley experiment. It established that the laws of physics are the same for all observers, regardless of their relative motion. This led to the groundbreaking concepts of time dilation (time running slower for objects in motion) and length contraction (objects appearing shorter in the direction of their motion).

1.13 - General Relativity:

Building upon special relativity, Einstein spent a decade grappling with the nature of gravity. In 1915, he unveiled his theory of general relativity, describing gravity not as a force, but as a curvature of spacetime caused by the presence of mass and energy. This revolutionary concept offered a more accurate explanation of gravity, particularly for phenomena like the bending of light in strong gravitational fields.

1.14 - E=mc²: A Symbol of Genius

The equation $E=mc^2$, often hailed as the most famous equation in physics, encapsulates the relationship between mass and energy. This simple yet profound equation, a consequence of special relativity, revealed that mass and energy are equivalent and can be converted into each other under certain conditions. This concept laid the foundation for the development of nuclear weapons, demonstrating the immense destructive potential harnessed within the atom. However, it also paved the way for advancements in nuclear power generation, highlighting the dual-edged nature of scientific discoveries.

1.15 - The Photoelectric Effect and the Birth of Quantum Mechanics

In 1905, the same year he published special relativity, Einstein tackled the puzzling phenomenon of the photoelectric effect. This phenomenon observed that light could eject electrons from certain materials, and the number of ejected electrons depended on the frequency of the light, not its intensity. By postulating that light exists as discrete packets of energy called photons, Einstein provided a revolutionary explanation for this effect. This concept, although initially met with skepticism, laid the groundwork for the development of quantum mechanics, one of the pillars of modern physics.

Einstein's work continues to inspire awe and influence scientific endeavors across diverse fields. His theories have provided the framework for numerous technological advancements, from GPS navigation to the development of new materials. However, his legacy extends beyond the realm of science, prompting profound philosophical questions about space, time, and the nature of reality. As we delve deeper into the historical context that shaped his genius, we gain a richer appreciation for the revolutionary impact of Albert Einstein and his enduring legacy on the human mind and our understanding of the universe.

1.2 - The Historical Context of Early 20th-Century Physics

To fully grasp the revolutionary nature of Einstein's work, we must transport ourselves back to the early 20th century and examine the existing scientific landscape. This period, often referred to as the "golden age of physics," witnessed a flurry of intellectual ferment and groundbreaking discoveries that set the stage for Einstein's groundbreaking theories.

1.21 The Prevailing Newtonian Framework:

For centuries, Isaac Newton's laws of motion and universal gravitation had dominated the scientific understanding of the universe. Newton's framework provided a robust and seemingly comprehensive explanation for the behavior of objects on both large (celestial) and small (terrestrial) scales. However, as scientific inquiry delved deeper, certain inconsistencies and limitations began to emerge.

- Challenges to Newtonian Mechanics: One such challenge arose from the Michelson-Morley experiment, conducted in 1887. This experiment aimed to detect the "ether," a hypothetical medium believed to be necessary for the propagation of light. However, the experiment consistently yielded a null result, suggesting that the ether either did not exist or was undetectable. This finding contradicted predictions based on Newtonian physics and opened the door for alternative explanations, paving the way for Einstein's theory of special relativity.

1.22 The Rise of New Scientific Ideas:

The late 19th and early 20th centuries witnessed a surge in scientific exploration and discovery, laying the groundwork for Einstein's revolutionary theories.

- Electromagnetism: The development of Maxwell's equations, which unified the theories of electricity and magnetism, provided a powerful framework for understanding the behavior of light and electromagnetism. However, these equations did not neatly fit within the existing Newtonian framework, hinting at the need for a more comprehensive theory.

- The Birth of Quantum Theory: The discovery of phenomena like blackbody radiation and the photoelectric effect challenged the classical understanding of light and matter. These discoveries ultimately led to the development of quantum mechanics, a new theory that explained the behavior of matter and energy at the atomic and subatomic levels. Although Einstein initially disagreed with some aspects of quantum mechanics, his work on the photoelectric effect played a crucial role in its development.

1.23 The Impact of Technology:

Technological advancements played a crucial role in both propelling scientific inquiry and providing crucial data for Einstein's theories.

- Michelson-Morley Experiment: As mentioned earlier, the null result of this experiment, made possible by advancements in optical technology, directly challenged the existing assumptions of Newtonian mechanics and paved the way for Einstein's theory of special relativity.

- Development of New Observational Tools: The invention of new telescopes and spectroscopic techniques allowed scientists to study celestial objects in greater detail. These observations, particularly regarding the behavior of light in strong gravitational fields, provided crucial evidence supporting the predictions of general relativity.

By understanding the historical context and the intellectual ferment of the early 20th century, we gain a deeper appreciation for the revolutionary nature of Einstein's work. His theories not only addressed the inconsistencies and limitations of the prevailing Newtonian framework but also emerged alongside other groundbreaking discoveries, marking a pivotal point in the history of physics and our understanding of the universe.

Chapter 2 - Prelude to Relativity
2.1 - Newtonian Mechanics and Its Limitations

Newtonian mechanics, formulated by Sir Isaac Newton in the 17th century, laid the foundation for classical physics and became a cornerstone of scientific understanding for centuries. This theory describes the motion of objects based on three fundamental laws: the law of inertia, the relationship between force and acceleration, and the principle of action and reaction. Despite its success in explaining the motion of everyday objects on Earth, Newtonian mechanics has its limitations, particularly when dealing with phenomena at extremely small scales, very high speeds, or strong gravitational fields.

2.11. Inadequacy at Quantum Scales:

Newtonian mechanics breaks down when dealing with the behavior of particles at the quantum level. At the subatomic scale, the principles of quantum mechanics take precedence, introducing concepts such as wave-particle duality, uncertainty, and quantum tunneling. Newtonian mechanics cannot accurately describe the probabilistic nature and discrete energy levels of particles observed in quantum systems.

2.12. Violation at Relativistic Speeds:

As objects approach the speed of light, Newtonian mechanics fails to accurately predict their behavior. Albert Einstein's theory of special relativity, developed in the early 20th century, demonstrated that as objects approach the speed of light, time dilation, length contraction, and mass increase occur. These relativistic effects are not accounted for in Newtonian mechanics, making it unsuitable for describing phenomena at near-light speeds.

3. Limited in Strong Gravitational Fields:

Newtonian gravity works well for objects with relatively small masses and at moderate distances. However, in situations where gravitational fields are extremely strong, such as near massive celestial bodies like black holes, the predictions made by Newton's law of gravitation deviate significantly from what is observed. General relativity, another theory formulated by Einstein, provides a more accurate description of gravity in such extreme conditions.

4. Cannot Explain Dark Matter and Dark Energy:

Newtonian mechanics and classical physics do not provide explanations for the observed phenomena of dark matter and dark energy. These substances, which together make up about 95% of the total mass-energy content of the universe, exhibit gravitational effects but do not interact with electromagnetic forces. Modern physics, including theories like general relativity and quantum field theory, attempts to address these mysteries.

5. Deterministic Nature vs. Quantum Indeterminacy:

Newtonian mechanics adheres to a deterministic worldview, where the future state of a system can, in theory, be precisely predicted given its initial conditions. However, quantum mechanics introduces inherent indeterminacy and probability into the fundamental nature of particles, challenging the determinism inherent in Newtonian physics.

Despite these limitations, Newtonian mechanics remains highly valuable and applicable in many practical situations, providing accurate predictions for a wide range of everyday phenomena. It serves as an excellent approximation for most macroscopic systems and continues to be an essential part of classical physics education. The development of quantum mechanics and relativity has expanded our understanding of the universe, revealing the need for more nuanced and comprehensive theories in different physical regimes.

2.1 - Maxwell's Equations and the Unification of Electricity and Magnetism

In the 19th century, James Clerk Maxwell revolutionized physics by unifying the previously separate theories of electricity and magnetism. His groundbreaking work, culminating in Maxwell's equations, demonstrated that these forces were two aspects of a single fundamental phenomenon – the electromagnetic field.

Maxwell's equations had profound implications for our understanding of the universe:

- **Electromagnetic waves:** The equations predicted the existence of electromagnetic waves, which could travel through empty space at a constant speed – the speed of light. This prediction was later experimentally confirmed, leading to the development of various technologies like radio, television, and radar.
- **The challenge to Newtonian mechanics:** The constant speed of light predicted by Maxwell's equations posed a challenge to Newtonian mechanics, as it implied that the speed of light was independent of the motion of the observer, which contradicted the Galilean transformations used in Newtonian mechanics.

This fundamental inconsistency between the established theory of Newtonian mechanics and the newly established properties of light, as described by Maxwell's equations, became a crucial impetus for the development of the theory of relativity

---

Chapter 3 - Special Relativity

Special relativity, a revolutionary theory conceived by Albert Einstein in 1905, stands as a cornerstone of modern physics. It challenged our classical understanding of space, time, and motion, painting a picture where these fundamental concepts are intertwined and relative, not absolute. This document delves deeper into the core aspects of special relativity, exploring the historical context, foundational principles, and its profound implications

3.1 - Michelson-Morley Experiment

Prior to the dawn of special relativity, the scientific community believed in the existence of a mysterious medium called the "aether," through which light waves were thought to propagate. The Michelson-Morley experiment aimed to detect the Earth's motion through this aether by measuring the subtle changes in the speed of light in different directions. However, the experiment delivered a surprising and unexpected null result – the speed of light remained constant regardless of the Earth's motion. This seemingly straightforward outcome had profound consequences, leaving the aether theory in shambles and paving the way for a new understanding of the universe.

3.2. The Pillars of the Theory: Einstein's Postulates

Special relativity rests firmly on two fundamental postulates:

1. The Principle of Relativity: This principle states that the laws of physics are the same for all observers in uniform motion (inertial frames of reference) relative to one another. In simpler terms, regardless of your constant speed and direction, the fundamental rules governing the universe remain consistent. Imagine two spaceships traveling at a constant velocity relative to each other. According to the principle of relativity, neither observer can claim to be "at rest" or possess a privileged frame of reference. The laws of physics, such as Newton's laws of motion, hold true for both observers in their respective frames.

2. The Constancy of the Speed of Light: This postulate states that the speed of light in a vacuum is the same for all observers, regardless of the motion of the light source or the observer itself. This seemingly simple statement has profound implications, contradicting our classical intuition about how velocities add up. For instance, if you are driving a car at 60 mph and turn on your headlights, from your perspective, the light appears to travel at its usual speed (around 3×10^8 m/s). According to the constancy postulate, an observer outside your car, even if moving in the opposite direction at 50 mph, would also measure the speed of light to be the same (3×10^8 m/s). This counterintuitive consequence arises from the inherent connection between space and time, a concept that classical mechanics failed to capture.

These seemingly simple postulates, when combined with the mathematical framework of Lorentz transformations, lead to a plethora of remarkable phenomena that defy our everyday experiences.

3.3. The Tapestry of Spacetime: Unveiling Time Dilation and Length Contraction

One of the most captivating consequences of special relativity is the phenomenon of time dilation. According to this concept, a moving clock appears to run slower compared to a stationary clock from the perspective of a stationary observer. The faster the object moves, relative to the observer, the greater the time dilation effect. This seemingly paradoxical phenomenon has been experimentally verified in numerous ways, including the observation of

----

high-energy particles called muons, which decay much slower than expected due to their high velocities relative to the Earth.

Length contraction is another counterintuitive consequence of special relativity. When an object moves at a significant fraction of the speed of light, its length appears to contract in the direction of its motion from the perspective of a stationary observer. This means that a moving object, like a spaceship, would appear shorter to a stationary observer compared to its true length when measured at rest.

These seemingly bizarre phenomena stem from the fundamental relationship between space and time in special relativity. The concept of absolute space and time, prevalent in classical mechanics, is replaced by a unified concept called spacetime. Events in spacetime are not absolute but are relative to the observer's frame of reference. The postulates of special relativity, combined with the mathematics of Lorentz transformations, dictate how these relationships between space and time change for different observers.

3.3 E=mc^2: The Equivalence of Mass and Energy:

Perhaps the most famous equation in physics, E=mc^2, is a direct consequence of special relativity. It expresses the equivalence of mass (m) and energy (E), stating that they are different forms of the same entity. This equation implies that even a small amount of mass can be converted into a tremendous amount of energy, as seen in nuclear reactions.

Special relativity has had a profound impact on our understanding of the universe and has numerous applications in various fields, including particle physics, cosmology, and GPS technology.

Chapter 4 - Experimental Confirmations

While the postulates of special relativity laid the theoretical groundwork, the theory's validity ultimately rests on its agreement with experimental observations. Over the past century, numerous experiments have confirmed the predictions of special relativity with remarkable accuracy, solidifying its position as a cornerstone of modern physics. This section delves into some key experimental confirmations, highlighting the ingenuity and meticulousness of scientists in validating this groundbreaking theory.

4.1 - Eddington's Eclipse Expedition (1919)

One of the first major confirmations of special relativity's implications came from the observation of the deflection of light by gravity. According to general relativity, a more comprehensive theory encompassing gravity, massive objects like the Sun warp the fabric of spacetime, causing light to bend as it passes near them. In 1915, Albert Einstein predicted that light rays passing close to the Sun would be deflected by an angle of about 1.75 arcseconds. However, directly observing this tiny deflection under normal circumstances would be impossible due to the Sun's overwhelming brightness.

The solution arrived in the form of a solar eclipse. During a total solar eclipse, the Sun's corona, its outer faint atmosphere, becomes briefly visible. This opportunity allowed Arthur Eddington, a British astronomer, to lead expeditions to Brazil and Africa in 1919 to photograph the stars near the eclipsed Sun. Comparing these

----

photographs with ones taken months later, when the Sun was not in the vicinity of the stars, revealed a measurable deflection in their apparent positions, consistent with Einstein's prediction. This observation provided strong evidence for the validity of general relativity, which incorporates special relativity as a special case when gravity is negligible.

4.2 - Time Dilation Experiments

The phenomenon of time dilation, a central prediction of special relativity, states that moving clocks run slower compared to stationary ones. While seemingly paradoxical, this prediction has been experimentally verified in numerous ways.

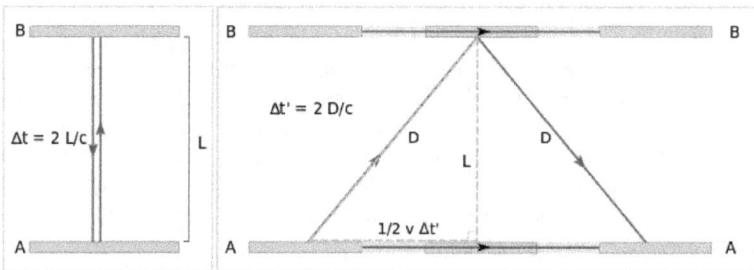

One compelling example involves **muons**, elementary particles created in the upper atmosphere by cosmic rays colliding with air molecules. Muons have a short intrinsic lifetime of about 2.2 microseconds. However, when measured on Earth, they travel considerable distances before decaying, seemingly defying their inherent lifespan. This apparent discrepancy can be explained by time dilation. Due to their high velocity (approaching the speed of light) relative to the Earth's surface, time runs slower for muons, effectively extending their lifespan from their perspective. This phenomenon has been experimentally confirmed with high precision.

Another experiment demonstrating time dilation involved utilizing highly accurate atomic clocks. These clocks were flown on airplanes or placed on satellites orbiting the Earth at high speeds. Upon comparing them with stationary clocks on the ground, a measurable time difference was observed, consistent with the predictions of special relativity.

4.2 - Particle Accelerators and Relativistic Effects

Particle accelerators are powerful machines that propel charged particles to extremely high velocities approaching the speed of light. These high velocities necessitate the application of special relativity to understand the behavior of these particles accurately.

One crucial consequence is the relativistic mass increase. As particles approach the speed of light, their mass appears to increase from the perspective of a stationary observer. This effect is crucial for calculating the energy required to accelerate particles to such high velocities in particle accelerators.

Furthermore, special relativity plays a vital role in understanding the decay times of unstable particles produced in these high-energy collisions. Time dilation dictates that these particles live longer from their perspective due to their high velocities, and this effect needs to be taken into account when analyzing their decay properties and lifetimes.

Experiments conducted at various particle accelerators, including the Large Hadron Collider (LHC), have consistently confirmed the predictions of special relativity. The behavior of high-energy particles, including their mass increase, decay times, and interactions, all align with the theoretical framework provided by special relativity.

Conclusion: A Symphony of Theory and Experiment

The experimental confirmations discussed here represent just a glimpse into the vast array of evidence supporting the validity of special relativity. From the deflection of light by gravity to the time dilation of muons and high-energy particle behavior, numerous experiments have consistently validated the predictions of this groundbreaking theory. This harmonious agreement between theory and experiment underscores the power of scientific inquiry and our quest to understand the fundamental nature of the universe. As we continue to explore the frontiers of physics, special relativity remains a vital tool, guiding us deeper into the remarkable tapestry of space, time, and the laws that govern them.

Chapter 5 - General Relativity

Special relativity, introduced earlier, revolutionized our understanding of space and time, revealing their intertwined nature. However, it left one crucial question unanswered: how does gravity, a fundamental force shaping the universe, fit into this new framework? This is where **general relativity**, another groundbreaking theory by Albert Einstein, takes center stage. Developed over a decade after special relativity, general relativity presents a comprehensive picture of gravity, replacing the classical notion of a force with a geometric interpretation of the universe.

5.1 - The Equivalence Principle

The foundation of general relativity lies in the **equivalence principle**. This principle, first proposed by Einstein, states that the effects of gravity are indistinguishable from the effects of being in an accelerating reference frame. In simpler terms, imagine being in a closed box, unable to see outside. If you experience a constant acceleration, it is impossible to differentiate whether this acceleration is due to external forces (like being in a rocket) or due to being in a gravitational field. This seemingly simple observation holds profound implications, suggesting that gravity is not a fundamental force but rather an emergent phenomenon arising from the underlying geometry of spacetime.

5.2 - Curved Spacetime and Gravity

General relativity proposes that the presence of mass and energy **distorts** the fabric of spacetime. This distortion is visualized as a curvature in the otherwise flat spacetime canvas. Imagine a bowling ball placed on a trampoline, causing it to sag. Similarly, massive objects like stars or planets warp the spacetime around them, creating a "gravity well." Objects moving in this curved spacetime follow the geodesics, the equivalent of straight lines in flat space, but appear to bend due to the curvature. This curvature is what we perceive as the force of gravity.

5.3. The Einstein Field Equations

The **Einstein field equations** are the heart of general relativity, a set of complex mathematical equations that describe the relationship between the curvature of spacetime and the distribution of mass and energy within it. These equations act as the bridge, connecting the geometric description of spacetime with the energy and momentum of matter. Solving these equations allows physicists to predict the behavior of objects in a gravitational field, including their trajectories, velocities, and even phenomena like the bending of light.

Implications and Applications of General Relativity: Beyond Theory

General relativity has revolutionized our understanding of gravity and has numerous profound implications across various scientific disciplines:

- **Black Holes:** These enigmatic objects, with their immense gravity, arise from the solutions of the Einstein field equations. Their existence, initially a theoretical prediction, has been confirmed through various observations, providing a glimpse into the remarkable consequences of extreme gravity.
- **Cosmology:** General relativity forms the foundation for our understanding of the large-scale structure and evolution of the universe. It allows scientists to study phenomena like the expansion of the universe, the formation of galaxies, and the behavior of dark matter and dark energy, which are believed to constitute most of the universe's energy content.
- **GPS Technology:** Even everyday technologies like GPS rely on the principles of general relativity. The gravitational time dilation experienced by GPS satellites orbiting the Earth needs to be taken into account for accurate positioning calculations.

5. Conclusion: A Universe Redefined

General relativity stands as a testament to the power of human ingenuity and our relentless pursuit of understanding the universe. It transcends the limitations of classical mechanics, offering a geometric interpretation of gravity that has profoundly impacted our perception of space, time, and the cosmos. From the mind-boggling implications of black holes to the vastness of the expanding universe, general relativity continues to guide our exploration of the universe's

----

mysteries, pushing the boundaries of scientific knowledge and inspiring further exploration.

Chapter 6 - Gravity's Bends: Predictions and Confirmations

Gravity, the invisible force that shapes the cosmos and keeps us grounded, has captivated and puzzled humans for centuries. While Isaac Newton's classical theory of gravity provided a powerful framework for understanding its effects, it did not fully explain its nature. Enter **general relativity**, Albert Einstein's groundbreaking theory that revolutionized our understanding of gravity, depicting it not as a force but as a curvature in the fabric of spacetime caused by the presence of mass and energy. This document explores some key predictions and confirmations of general relativity, highlighting how these "bends" in spacetime manifest in the universe around us.

6.1 - Gravitational Redshift

One of the first and most intriguing predictions of general relativity is **gravitational redshift**. This phenomenon states that light emitted from an object in a strong gravitational field (like a star) will experience a **shift towards the red end of the electromagnetic spectrum**. Imagine climbing out of a deep gravity well; just like it takes more energy to climb out, it also takes more energy for light to escape such a well. This loss of energy translates to a **decrease in frequency** and an **increase in wavelength** of the emitted light, pushing it towards the red end of the spectrum.

Confirmations:

- **Early Observations:** In the early 1920s, astronomers observed a consistent redshift in the spectral lines of light emitted from distant galaxies, initially interpreted as evidence of an expanding universe. Later, it was recognized that a portion of this redshift could be attributed to the gravitational effect of these massive galaxies.
- **Laboratory Experiments:** In the late 20th century, scientists conducted experiments using highly accurate atomic clocks. Placing one clock at a higher altitude than another effectively subjected it to a weaker gravitational field. After a period, a measurable difference in time was observed, with the clock in the higher altitude running slightly faster than the one at lower altitude. This confirmed the relationship between time and

---

gravity, which, when extrapolated, implies the existence of gravitational redshift.

6.2. The Deflection of Light

Another fascinating prediction of general relativity is the **deflection of light** by gravity. According to the theory, the presence of mass curves the fabric of spacetime, influencing the path of any object, including light, which travels along the geodesics of this curved space. This means that light passing near a massive object, like the Sun, will have its trajectory bent slightly due to the curvature of spacetime around it.

Confirmations:

- **Eddington's Eclipse Expedition (1919):** During a total solar eclipse in 1919, British astronomer Arthur Eddington led expeditions to photograph the positions of stars near the eclipsed Sun. Comparing these photographs with ones taken months later, when the Sun was not in the vicinity of the stars, revealed a measurable deflection in their apparent positions, consistent with Einstein's prediction. This observation provided strong evidence for the validity of general relativity.
- **Modern Gravitational Lens Observations:** Astronomers utilize a phenomenon called **gravitational lensing** to study distant objects in the universe. This phenomenon occurs when the light from a distant object, like a galaxy, is bent by the gravitational field of a massive object, like a galaxy cluster, acting as a lens. By analyzing the distorted images created by gravitational lensing, scientists can confirm the presence of massive objects and even probe the distribution of dark matter in the universe.

6.3. Time Dilation in a Gravitational Field

General relativity also predicts that **time runs slower in stronger gravitational fields**. This phenomenon, known as **gravitational time dilation**, is a consequence of the curvature of spacetime. Imagine time flowing like a river; in regions with strong gravity, the river flows slower compared to regions with weaker gravity.

Confirmations:

- **Atomic Clock Experiments:** Experiments using extremely precise atomic clocks have been conducted to confirm time dilation in Earth's gravitational field. Placing one clock at a higher altitude, where the gravitational field is weaker, and comparing it with a clock at a lower altitude has revealed a measurable difference in time, with the clock at higher altitude running slightly faster.
- **GPS Technology:** Even everyday technology like GPS relies on the principles of general relativity. Satellites orbiting the Earth experience time dilation due to their higher altitude and weaker gravitational field. This effect needs to be taken into account for accurate positioning calculations, ensuring the precise functioning of GPS systems.

Chapter 7 - Legacy and Ongoing Impact

General relativity, Albert Einstein's groundbreaking theory of gravity, stands as a cornerstone of modern physics, revolutionizing our understanding of the universe. Its impact extends far beyond theoretical elegance, influencing various scientific fields and even shaping practical applications we encounter daily. This document delves into the enduring legacy of general relativity, exploring its practical applications and the ongoing research efforts fueled by its profound insights.

7.1 - A Legacy of Profound Insights

General relativity's legacy is multifaceted, encompassing:

- **A New Understanding of Gravity**: It replaced the classical notion of gravity as a force with a geometric interpretation, depicting it as the curvature of spacetime caused by mass and energy. This shift in perspective has had profound implications for our understanding of the universe's fundamental structure and behavior.
- **A Gateway to Further Exploration**: It opened doors to exploring exotic phenomena like black holes, neutron stars, and the expanding universe. These once hypothetical concepts are now actively studied, with observations providing evidence for their existence and pushing the boundaries of our knowledge.
- **A Foundation for Modern Cosmology**: It forms the cornerstone of modern cosmology, providing the framework for understanding the large-scale structure and evolution of the universe. From the Big Bang to the accelerating expansion, general relativity plays a crucial role in deciphering the universe's history and future.

7.2. Practical Applications: From GPS to Gravitational Wave Detection

General relativity's influence transcends theoretical physics, impacting various practical applications:

- **GPS Technology:** The Global Positioning System (GPS) relies heavily on the principles of general relativity to function accurately. The time dilation experienced by GPS satellites orbiting the Earth, due to their higher altitude and weaker gravitational field, needs to be factored into calculations to ensure precise positioning.
- **Satellite Navigation Systems:** Similar to GPS, other satellite navigation systems like GLONASS (Russia) and Galileo (Europe) also incorporate relativistic corrections to account for the effects of gravity on their orbital paths and ensure accurate positioning data.
- **Gravitational Wave Detection:** The recent advancements in detecting gravitational waves, ripples in the fabric of spacetime predicted by general relativity, have opened a new window to studying the universe. These detections, achieved by sophisticated instruments like LIGO and Virgo, provide valuable insights into phenomena like black hole mergers and neutron star collisions.

7.3- Current Research in Gravitational Waves

The detection of gravitational waves has ignited a new era of gravitational-wave astronomy, with ongoing research focusing on:

- **Improving Sensitivity:** Scientists are constantly striving to improve the sensitivity of existing gravitational-wave detectors and develop new ones, aiming to observe a wider range of gravitational wave sources and capture fainter signals.
- **Expanding the Observational Window:** With advancements in technology, scientists are exploring the possibility of detecting gravitational waves at other frequencies beyond the current range, potentially revealing new types of sources and phenomena.
- **Multi-Messenger Astronomy:** Combining gravitational-wave observations with data from other telescopes and instruments across the electromagnetic spectrum allows for a more comprehensive understanding of astronomical events, providing a multifaceted view of the universe's most violent and energetic phenomena.

Conclusion: A Journey of Discovery Continues

General relativity's legacy continues to grow as scientists delve deeper into its implications and refine our understanding of the universe. From its practical applications in everyday technology to the ongoing exploration of gravitational waves, the theory serves as a powerful tool for unraveling the universe's mysteries. As research progresses, we can expect further breakthroughs and discoveries, pushing the boundaries of knowledge and revealing new facets of the remarkable universe we inhabit.

Chapter 8 - Einstein vs Quantum Mechanics

Albert Einstein, one of the architects of modern physics, revolutionized our understanding of space, time, and gravity with his theory of relativity. However, when it came to the emerging field of quantum mechanics, which describes the behavior of matter and energy at the atomic and subatomic level, his relationship with the theory became one of intellectual struggle. This section delves into the key points of contention between Einstein and quantum mechanics, focusing on the famous **Einstein-Podolsky-Rosen (EPR) paradox** and the **Bohr-Einstein debates**, culminating in the fascinating concept of **quantum entanglement and nonlocality**.

8.1 - The EPR Paradox

In 1935, Einstein, along with Boris Podolsky and Rosen, published a paper titled "Can Quantum-Mechanical Description of Reality Be Considered Complete?" This paper presented the **EPR paradox**, a thought experiment designed to highlight what they perceived as an inherent incompleteness in the prevailing interpretation of quantum mechanics, known as the Copenhagen interpretation.

The EPR Paradox:

Imagine a source emitting two entangled particles, such as electrons, with their spins (a property of the particle) correlated in a specific way. Suppose we measure the spin of one particle along a specific axis and find it to be "up." According to the Copenhagen interpretation, the other entangled particle instantly collapses into a state with "down" spin, regardless of the distance between them.

Einstein found this "spooky action at a distance" troubling. He argued that no physical signal could transmit information faster than the speed of light, a principle enshrined in his theory of relativity. Therefore, he challenged the notion that the measurement on one particle instantaneously "forces" the other particle to assume the opposite spin, regardless of their separation.

8.2 - The Bohr-Einstein Debates

The EPR paradox ignited a series of intellectual discussions and debates between Einstein and Niels Bohr, a prominent advocate for the Copenhagen interpretation of quantum mechanics. These debates highlighted the fundamental differences in their viewpoints:

- Einstein favored a local, deterministic view of reality. He believed that particles possess definite properties even when not measured and that any interactions between them must occur through local interactions, obeying the speed limit of light.
- Bohr emphasized the probabilistic nature of quantum mechanics. He argued that particles exist in a superposition of states until measured, and the act of measurement collapses the wavefunction, determining their properties. He also pointed out the principle of complementarity, where certain complementary properties, like position and momentum, cannot be known simultaneously with perfect accuracy.

While these debates never reached a definitive conclusion, they significantly advanced the understanding of the interpretations and limitations of quantum mechanics.

8.3 - The Bohr-Einstein Debates

Further experimentation and advancements in technology provided evidence for the phenomenon of **quantum entanglement**. This phenomenon, seemingly defying the principles of classical physics, describes the creation of pairs of particles in a way that their properties become inextricably linked, regardless of the distance separating them. Measuring the state of one entangled particle instantaneously determines the state of the other, in accordance with the predictions of the Copenhagen interpretation.

Experiments have successfully demonstrated entanglement over increasing distances, challenging Einstein's notion of locality. However, it is essential to understand that this **does not** involve faster-than-light communication. While the particles share a correlated state, they do not exchange information or influence each other directly. The correlation is established during the entanglement process itself, not through any ongoing interaction.

8.4 - Quantum Entanglement and Nonlocality

Further experimentation and advancements in technology provided evidence for the phenomenon of **quantum entanglement**. This phenomenon, seemingly defying the principles of classical physics, describes the creation of pairs of particles in a way that their properties become inextricably linked, regardless of the distance separating them. Measuring the state of one entangled particle instantaneously determines the state of the other, in accordance with the predictions of the Copenhagen interpretation.

Experiments have successfully demonstrated entanglement over increasing distances, challenging Einstein's notion of locality. However, it is essential to understand that this **does not** involve faster-than-light communication. While the particles share a correlated state, they do not exchange information or influence each other directly. The correlation is established during the entanglement process itself, not through any ongoing interaction.

Conclusion: A Legacy of Provocative Questions

Einstein's initial skepticism towards quantum mechanics, particularly the EPR paradox, played a crucial role in stimulating further research and refining our understanding of this fascinating theory. While he never fully embraced the seemingly "spooky" aspects of entanglement, his contributions highlighted the need for careful consideration of interpretations and potential limitations. The ongoing exploration of quantum mechanics, along with advancements in entanglement research, continues to push the boundaries of our knowledge and provoke new questions about the nature of reality and the limits of our current understanding.

Chapter 9 - Relativity in the 21st Century

Relativity, both special and general, remains a cornerstone of modern physics in the 21st century. Developed by Albert Einstein in the early 20th century, these theories revolutionized our understanding of space, time, gravity, and the universe. Today, these theories continue to be actively researched and tested, with new discoveries and advancements constantly pushing the boundaries of our knowledge.

9.1 - Advancements in Experimental Testing

Since its inception, relativity has undergone numerous experimental tests to validate its predictions. In the 21st century, the development of ever more sophisticated technologies has allowed scientists to conduct these tests with unprecedented precision:

- **GPS**: The Global Positioning System relies on a network of satellites orbiting Earth. These satellites experience time dilation due to their high velocity, which must be accounted for by relativity to ensure accurate positioning.
- **Particle accelerators**: Experiments at facilities like the Large Hadron Collider (LHC) test the predictions of relativity by colliding particles at near light speed and observing their behavior. These experiments have confirmed the existence of antimatter and the behavior of particles under extreme conditions, consistent with relativistic predictions.
- **Gravitational wave observations**: The 2015 detection of gravitational waves by the LIGO (Laser Interferometer Gravitational-Wave Observatory) marked a significant milestone in the history of relativity. These waves, +predicted by general relativity, carry information about the movement of massive objects in the universe, providing a new way to test the theory.

These advancements, along with numerous other ongoing experiments, continue to solidify the foundation of relativity and pave the way for further exploration of its implications.

9.2 - Modern Cosmological Discoveries

Relativity plays a crucial role in our understanding of the universe's origin and evolution. Cosmological observations in the 21st century have yielded fascinating discoveries that challenge and refine our current understanding:

- **Cosmic Microwave Background Radiation (CMB)**: The CMB is a faint afterglow of the Big Bang, carrying information about the universe's early stages. Precise measurements of the CMB by missions like WMAP (Wilkinson Microwave Anisotropy Probe) and Planck have provided crucial insights into the universe's age, composition, and expansion rate.
- **Dark matter and dark energy**: Observations suggest that the vast majority of the universe's matter and energy is in forms we cannot directly observe, dubbed dark matter and dark energy. While the nature of these remains a mystery, their existence is consistent with the predictions of general relativity on a large scale.
- **Accelerating expansion**: Recent observations suggest that the expansion of the universe is accelerating, potentially driven by dark energy. This discovery has significant implications for the future of the universe and is an active area of research in cosmology.

These discoveries highlight the ongoing interplay between relativity and cosmological observations. As we develop new telescopes and instruments, we can expect further advancements that will deepen our understanding of the universe and its evolution.

9.3 - The Role of Relativity in Unified Theories

One of the major goals of modern physics is to develop a unified theory that describes all the fundamental forces of nature, including gravity. While both special and general relativity have been remarkably successful in their respective domains, they remain incompatible with the framework of quantum mechanics, the theory governing the behavior of the universe at the atomic and subatomic level.

One promising approach to unification is **string theory**, which proposes that fundamental particles are not point-like but rather tiny, vibrating strings. String theory incorporates elements of both relativity and quantum mechanics, but it remains highly theoretical and requires further development and experimental verification.

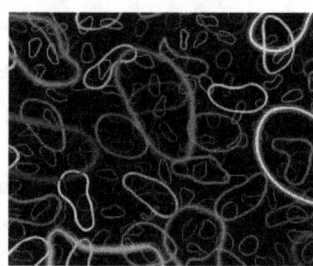

In conclusion, relativity remains a vibrant and evolving field in the 21st century. Advancements in experimental testing, modern cosmological discoveries, and the ongoing pursuit of a unified theory all contribute to our ever-deepening understanding of the universe and its fundamental laws.

Chapter 10 - Beyond Einstein

While general relativity has achieved remarkable success, it is not without its challenges:

10.1 - Challenges to General Relativity

- **Singularities**: General relativity predicts the existence of singularities, points of infinite density and curvature of spacetime, occurring at the center of black holes and the Big Bang. These singularities challenge our understanding of physics as our current physical models break down at these points.
- **Dark matter and dark energy**: As mentioned earlier, the existence of dark matter and dark energy, which are not explained by general relativity, suggests the need for a more comprehensive theory.
- **Quantum gravity**: General relativity is a classical theory, incompatible with the principles of quantum mechanics. This incompatibility hinders our ability to describe phenomena at the quantum level, such as the behavior of matter at extremely high densities or the Hawking radiation emitted from black holes.

These challenges motivate researchers to explore alternative theories and seek a more comprehensive understanding of gravity.

10.2 - Modified Gravity Theories

In an attempt to address the limitations of general relativity, several modified gravity theories have been proposed:

- **f(R) gravity**: This theory modifies the gravitational field equation by adding a function of the Ricci scalar (R), a measure of spacetime curvature. This allows for the inclusion of dark energy and potentially avoids singularities.
- **Massive gravity**: This theory introduces a small mass to the graviton, the hypothetical particle carrying the force of gravity. This modification can address certain challenges of general relativity at the quantum level.

These are just a few examples, and many other modified gravity theories are actively being explored.

10.3 - The Quest for a Theory of Everything

The ultimate goal of many physicists is to develop a "Theory of Everything" (TOE) that would unify all the fundamental forces of nature (gravity, electromagnetism, strong nuclear force, and weak nuclear force) within a single theoretical framework. String theory, mentioned earlier, is a promising candidate for a TOE as it incorporates gravity and attempts to reconcile it with quantum mechanics. However, string theory remains highly theoretical and requires further development and experimental verification.

The journey beyond Einstein is a continuous exploration of the nature of gravity and its role in the universe. While general relativity remains a cornerstone of modern physics, the quest for a more comprehensive understanding demands us to explore alternative theories and push the boundaries of our current knowledge. As we delve deeper into the mysteries of gravity, we may unlock new insights into the universe and its fundamental laws.

Conclusion

Einstein's Enduring Legacy:

Albert Einstein's revolutionary theories of relativity continue to hold immense significance in the 21st century. His work not only transformed our understanding of space, time, and gravity but also laid the foundation for further exploration of the universe.

The enduring legacy of Einstein extends beyond his scientific contributions. He is also remembered for his pursuit of peace, his humanitarian efforts, and his unwavering curiosity about the universe.

Future Prospects and Unanswered Questions:

While relativity has achieved remarkable success, there are still many unanswered questions and ongoing challenges that motivate researchers to delve deeper:

- Unifying gravity with quantum mechanics: Finding a theory that reconciles these two pillars of physics is crucial for understanding the behavior of matter at extreme conditions and the nature of singularities.
- Understanding dark matter and dark energy: Elucidating the nature of these mysterious components of the universe remains a major challenge, potentially requiring new theories that go beyond general relativity.
- The search for a Theory of Everything: The quest for a unified theory that encompasses all fundamental forces and phenomena continues to be a driving force in theoretical physics.

The future of gravity research holds immense promise. As we develop new technologies, conduct advanced experiments, and refine theoretical frameworks, we can expect exciting discoveries that will deepen our understanding of the universe and its fundamental laws.

In conclusion, Einstein's legacy remains a source of inspiration and a starting point for further exploration. As we continue to explore the mysteries of gravity and the universe, we honor his dedication to scientific inquiry and his pursuit of a deeper understanding of the cosmos.

www.ingramcontent.com/pod-product-compliance
Lightning Source LLC
Chambersburg PA
CBHW071018290526
45795CB00005B/1848